Numerical Adventures:

Explore, Learn and Have Fun

Scan to receive free information and resources

You can also write to
mariledys@educkidsonline.com

Welcome to "Numerical Adventures"!

Get ready to embark on an exciting journey full of exploration, learning and fun in the world of mathematics! In this book, children will discover a fascinating universe of numbers that will allow them to develop their math skills while having fun.

Explore the Secrets of the Numerical World:

- **Addition and Subtraction:** Let's exercise these 5- and 6-digit arithmetic operations with different approaches that will help you solve math problems and everyday situations.
- **Multiplication and Division:** Discover how multiplication and division can help us solve real-world problems. You will learn how to use these operations in creative and exciting ways.
- **Place Value:** We will discover how the place a digit occupies in a number determines its value. We will learn to identify and understand the value of each position, from ones, tens, hundreds and more.
- **Fractions:** Dive into the world of fractions as you explore how they relate to everyday situations and number challenges. Ready to divide a pizza into equal parts or split a chocolate bar among friends?
- **Capacity Measurements:** Discover how to measure different quantities and capacities in fun, practical situations - how much water can a swimming pool hold, and how much food do we need for a party?
- **Coordinate Planes:** Master coordinate planes as you solve mysteries and find hidden treasures - use your code-cracking skills to find the right path to victory!
- **Units of Measurement:** we will explore the units of measurement that help us quantify length, mass and capacity. Through hands-on activities and everyday examples, children will learn to apply these units of measure in real situations, developing essential skills for daily life and problem solving.

- Ordinal Numbers: They help us to order objects or events in a sequence. We will learn to identify and use ordinal numbers. We will discover how they are written and pronounced, as well as their importance in everyday life. In addition, we will have fun with interactive activities and hands-on exercises to reinforce this fundamental concept in mathematics.

Join the excitement of "Numerical Adventures"!

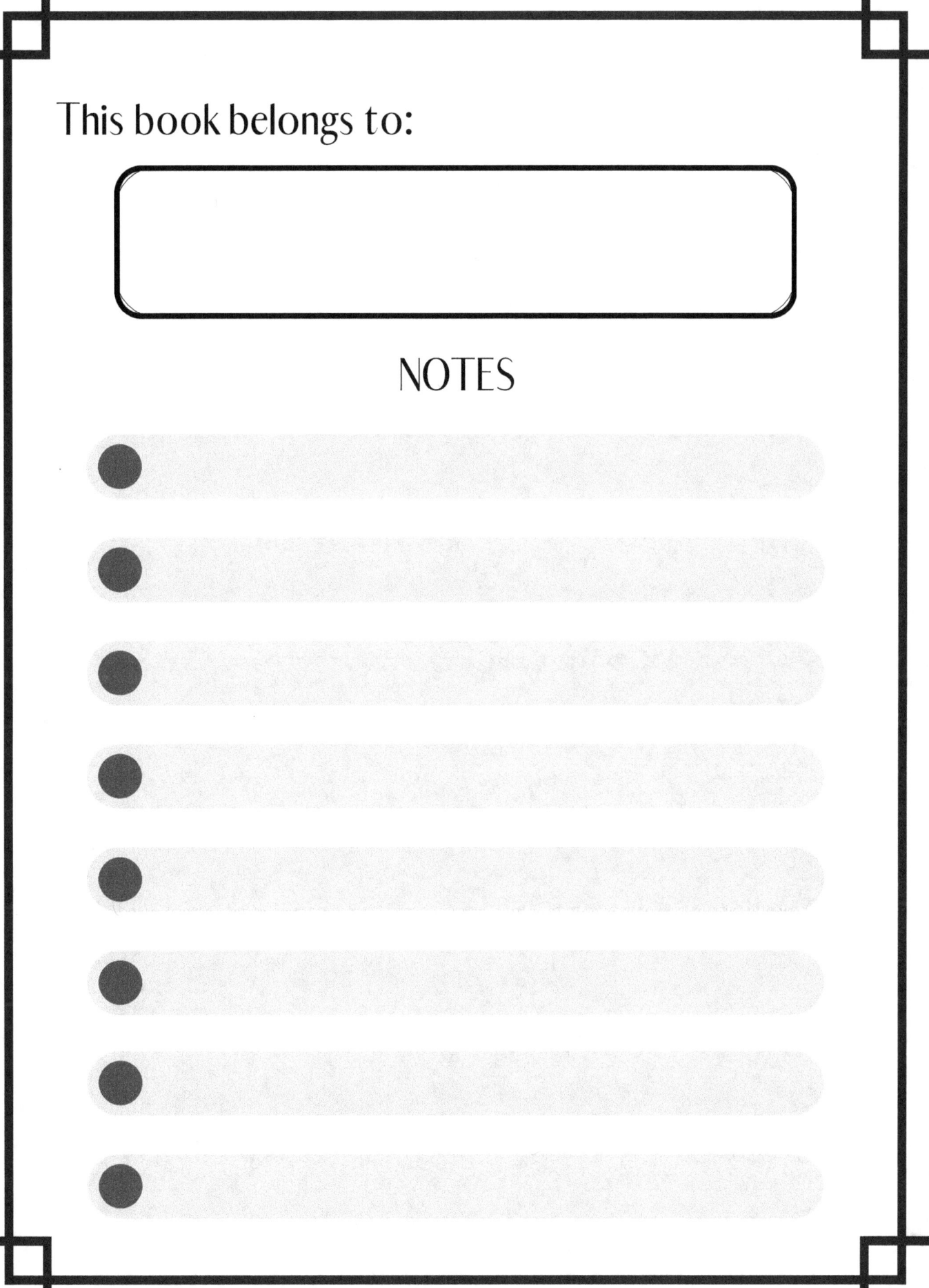

INDEX

- **9** — AMOUNT
- **15** — REMAINS
- **19** — PLACE VALUE
- **26** — MULTIPLICATION
- **33** — DIVISION
- **40** — FRACTIONS
- **46** — COORDINATE PLANE

INDEX

49 UNITS OF MEASUREMENT

72 ORDINAL NUMBERS

ADDING BIG NUMBERS

Solve 5-digit sums

16.573	25.903	36.885	58.944	45.322 +
14,928	+ 75.859	+ 57.573	+ 45.943	+ 54.264

74.932	83.930	13,213	14.392	43.242
74.931	+ 25.632	+ 21,937	+ 15.883	+ 39.224

53.246	29,322	12,298	69.392	38.983
42.038	+ 33.993	39.839	+ 34.827	+ 13,223

35.292	13.938	20.938	56.388	46.892
17.202	+ 14,992	+ 13.993	+ 33.992	+ 29.023

ADDING BIG NUMBERS

Solve 6-digit sums

| 854,354 | 754,834 | 938,575 | 208,422 | 350,482 |
| + 255,681 | + 759,387 | + 398,756 | + 627,520 | + 962,908 |

| 529,045 | 658,392 | 568,904 | 890,523 | 782,433 |
| + 986,023 | + 423,578 | + 427,835 | + 543,809 | + 264,897 |

| 348,794 | 897,945 | 920,735 | 429,824 | 782,433 |
| + 348,952 | + 837,234 | + 624,875 | + 576,234 | + 649,346 |

| 984,629 | 649,243 | 982,492 | 846,483 | 778,333 |
| + 584,209 | + 958,933 | + 774,928 | + 232,453 | + 889,235 |

ADDING BIG NUMBERS

Find the missing number in each problem.

529.045 + _____ = 986.023   423,578 + _____ = 427.835

76.489 + _____ = 84.790   357.865 + _____ = 125.969

575.852 + _____ = 753.985   247,975 + _____ = 357,975

542,986 + _____ = 927.835   120.839 + _____ 567,832

305.895 + _____ = 835.453   423,578 + _____ = 427.835

ADDING BIG NUMBERS

Solve each operation.

```
  12.353      65.722     124,244      96,972       7.532
   4,324       7.533       2,343     898.800       3.421
+ 42.632    + 54.322    + 60.033    + 24.642    + 12,392
_____    _____    _____    _____    _____


  97.584     756.28     275,698      87.963       2.995
  17.067     12.1803      1.742     786.452         675
+ 25.632    +  6.548   +    654    + 765.412    + 76.451
_____    _____    _____    _____    _____


   7.890     98.010     634.570      98.986       8.765
  32.416     52.674       5.876      65.890         563
+ 25.212    +  7.145    + 39.012    + 13.523    + 11.356
_____    _____    _____    _____    _____
```

ADDING WHEEL

Find and color the sum of the central number. Look at the example

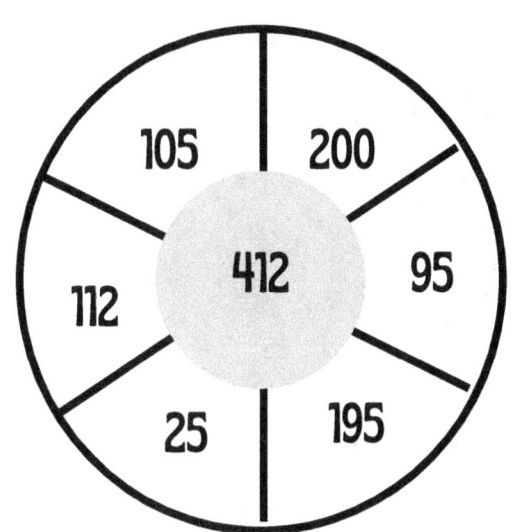
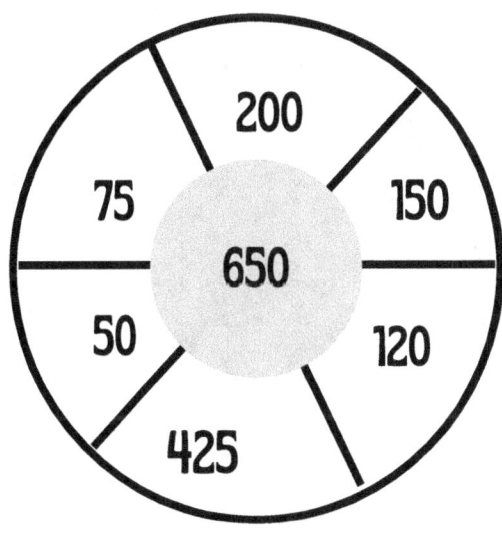
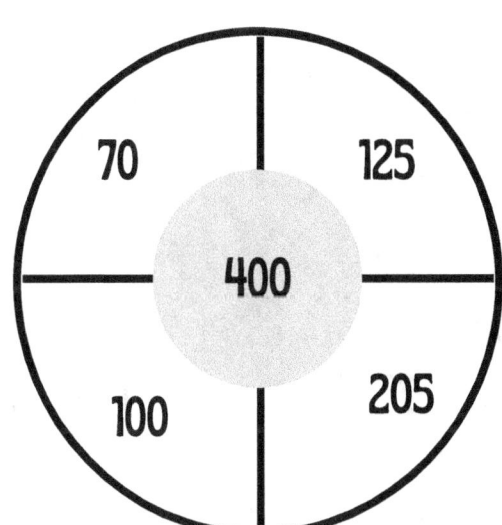

ADDING WHEEL

Find and color the sum of the central number

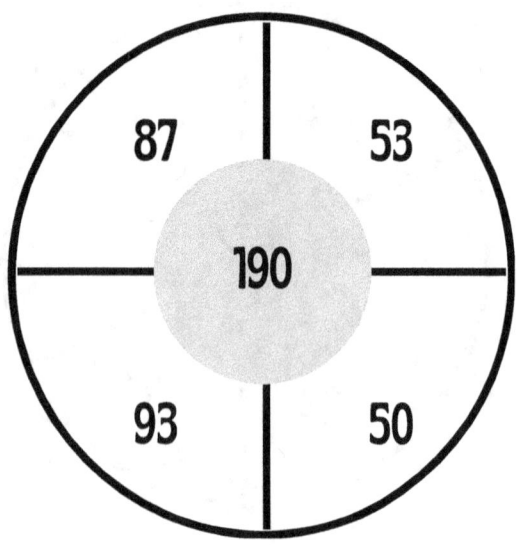

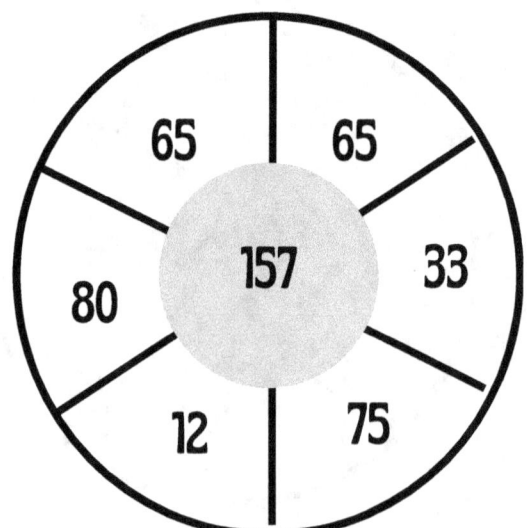

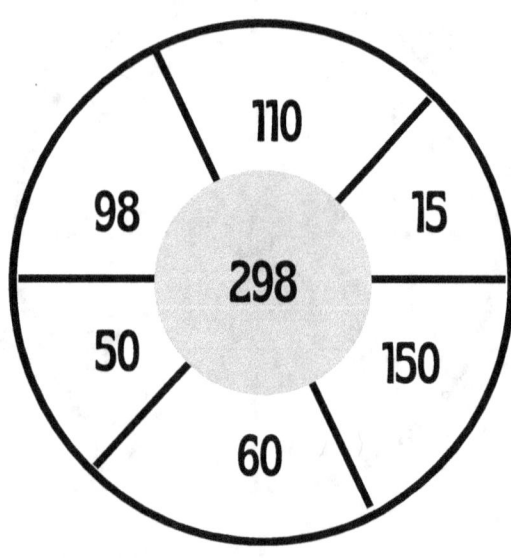

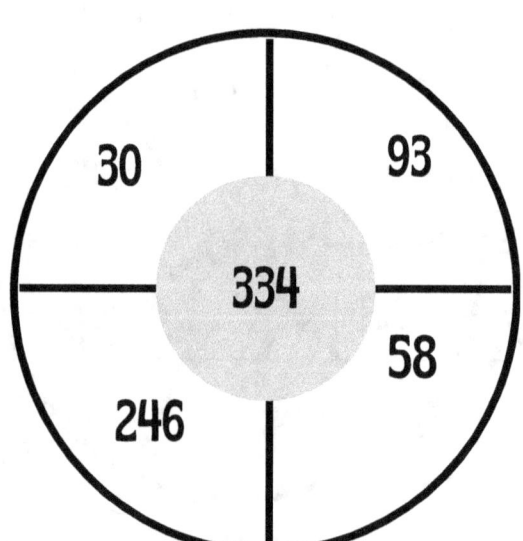

SUBTRACT LARGE NUMBERS

Solve 6-digit subtractions

854.354 - 354.512	973.952 - 759.387	938.575 - 398,756	471.358 - 241.906	897.453 - 753.927
865.865 - 354.187	753.498 - 538.098	418.342 - 356.317	890.523 - 543.809	782.533 - 264.897
964.154 - 348.952	897.945 - 837.234	920.735 - 624.875	654.159 - 576.234	782.433 - 149.346
984.629 - 584.209	531.035 - 345.931	941.427 - 774.928	846.483 - 232,453	542.679 - 495.274

SUBTRACT LARGE NUMBERS

Solve 6-digit subtractions

| 357.925 | 658.947 | 645.285 | 423.637 | 905.753 |
| - 257.374 | - 437.936 | - 426.482 | - 179.538 | - 876.132 |

| 530.725 | 835.516 | 178.954 | 962.543 | 645.297 |
| - 56.745 | - 643.486 | - 67.834 | - 486.259 | - 342.368 |

| 564.296 | 756.476 | 846.638 | 538.498 | 819.649 |
| - 248.576 | - 543.253 | - 436.387 | - 327.590 | - 365.938 |

| 698.078 | 698.408 | 864.774 | 923.387 | 456.234 |
| - 603.756 | - 524.264 | - 356.928 | - 312.453 | - 236.135 |

SUBTRACTION WHEEL

Complete the wheel with the results of the subtractions

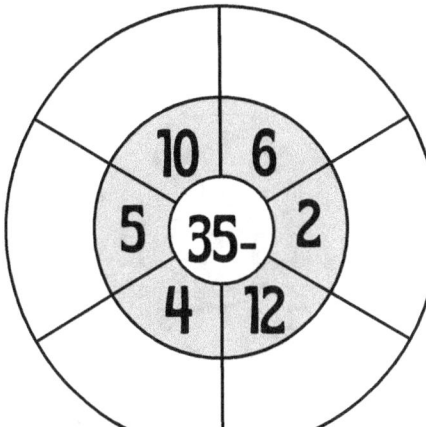
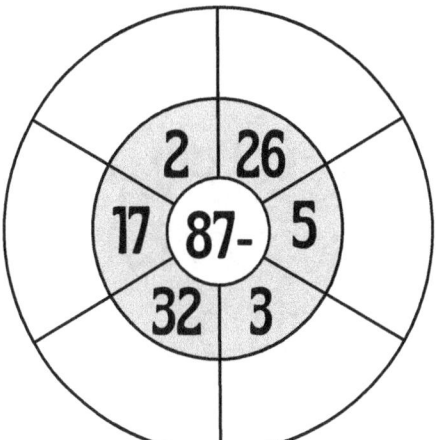
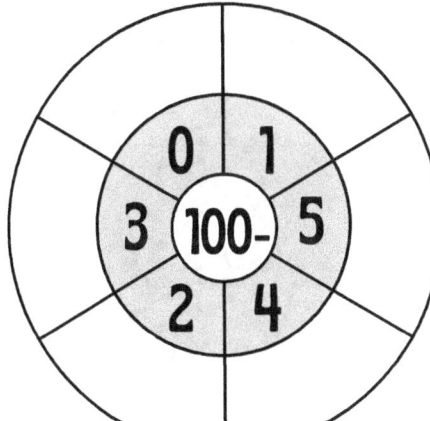
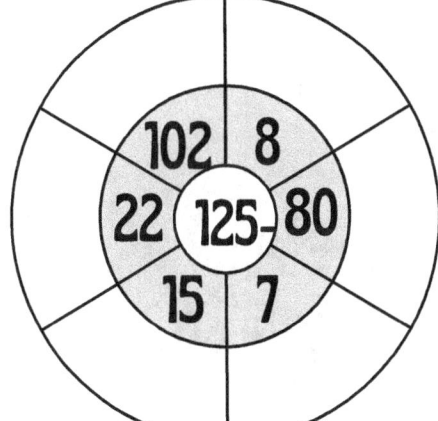
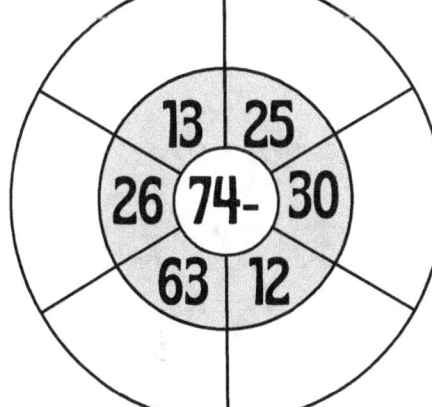
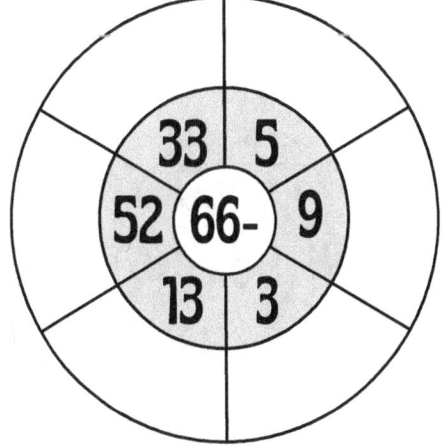

SUBTRACTION WHEEL

Complete the wheel with the results of the subtractions

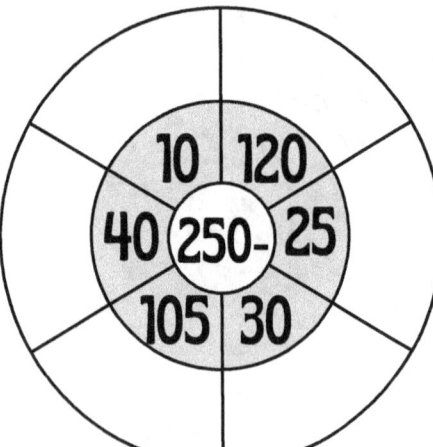
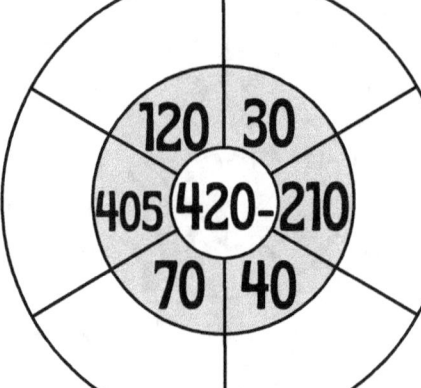
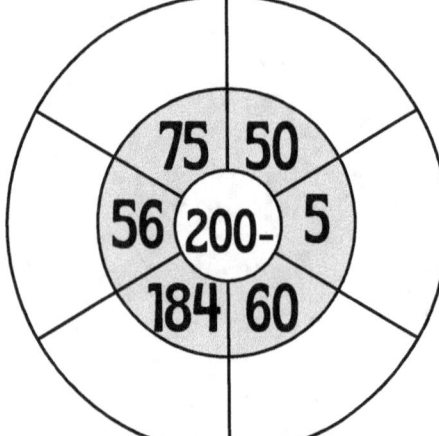

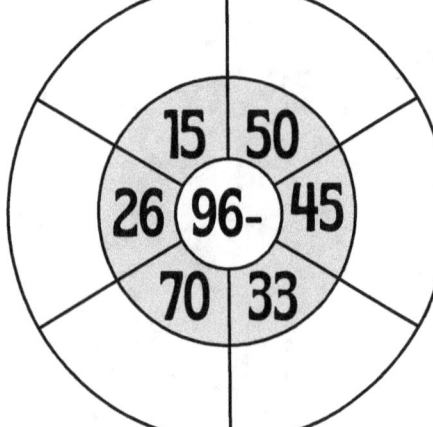

PLACE VALUE

Place each figure in its position

466

Hundreds	dozens	Units

854

Hundreds	dozens	Units

705

Hundreds	dozens	Units

86

Hundreds	dozens	Units

127

Hundreds	dozens	Units

PLACE VALUE

Place each figure in its position

1.025

Thousand Units	Hundreds	dozens	Units

4.379

Thousand Units	Hundreds	dozens	Units

4.503

Thousand Units	Hundreds	dozens	Units

8.948

Thousand Units	Hundreds	dozens	Units

PLACE VALUE

Place each figure in its position

54.754

Tens of Thousand	Thousand Units	Hundreds	dozens	Units

60.295

Tens of Thousand	Thousand Units	Hundreds	dozens	Units

31.290

Tens of Thousand	Thousand Units	Hundreds	dozens	Units

23.087

Tens of Thousand	Thousand Units	Hundreds	dozens	Units

PLACE VALUE

Place each figure in its position

650.834

Hundreds of Thousands	Tens of Thousand	Thousand Units	Hundreds	dozens	Units

532.054

Hundreds of Thousands	Tens of Thousand	Thousand Units	Hundreds	dozens	Units

769.251

Hundreds of Thousands	Tens of Thousand	Thousand Units	Hundreds	dozens	Units

942.692

Hundreds of Thousands	Tens of Thousand	Thousand Units	Hundreds	dozens	Units

PLACE VALUE

Mark the correct answer with a cross.

What number is made up of...?

      7 Units
      3 Dozens
      4 Hundreds
      8 Thousand Units

(7.348)   (8.437)   (8.473)   (7.384)

What number is made up of...?

      9 Dozens
      1 Millar Unit
      9 Hundreds
      2 Units

(9.192)   (1.992)   (9.291)   (1.299)

PLACE VALUE

Mark the correct answer with a cross.

What number is made up of...?

2 Dozens
2 Thousand units
2 Units

(2.202) (2.220) (222) (2.022)

What number is made up of...?

4 Hundreds
2 Units
9 Dozens
4 Thousand Units

(4.294) (9.424) (4.492) (2.944)

PLACE VALUE

Mark the correct answer with a cross.

What number is made up of 9 C, 5D and 3U?

(539) (953) (395) (9.053)

What number is made up of 7C and 8U?

(780) (78) (708) (807)

What number is made up of 2UM, 5C, 6D and 5U?

(6.552) (2.565) (2.655) (5.652)

What number is made up of 5D and 2U?

(25) (520) (52) (502)

What number is made up of 6U, 4D, 9C?

(464) (964) (9046) (946)

MULTIPLICATION: DECENS, HUNDREDS, UNITS OF THOUSANDS

Color the box with the correct answer

The result of 9 x 10 is...

(99)   (90)   (190)   (109)

The result of 86 x 10 is...

(806)   (860)   (8.600)   (86)

The result of 75 x 100 is...

(750)   (7.050)   (7.500)   (75.000)

The result of 25 x 100 is...

(2.500)   (250)   (125)   (25.000)

MULTIPLICATION: DECENS, HUNDREDS, UNITS OF THOUSANDS

Color the box with the correct answer

The result of 225 x 10 is...

(225) (2.225) (1.225) (2.250)

The result of 70 x 1,000 is...

(7.000) (70.000) (70.100) (700)

The result of 75 x 1,000 is...

(750) (7.050) (7.500) (75.000)

The result of 96 x 1,000 is...

(900) (96.000) (9.600) (96.100)

MULTIPLICATION: DECENS, HUNDREDS, UNITS OF THOUSANDS

Solve the following multiplications

- 25 x 100 =
- 242 x 100 =
- 45 x 100 =
- 310 x 100 =
- 2035 x 100 =
- 345 x 100 =
- 100 x 100 =
- 1348 x 100 =

- 54 x 1000 =
- 75 x 1000 =
- 726 x 1000 =
- 2150 x 1000 =
- 853 x 1000 =
- 328 x 1000 =
- 250 x 1000 =
- 3954 x 1000 =

MULTIPLYING WHEEL

Complete the wheel. Multiplication from 1 to 9

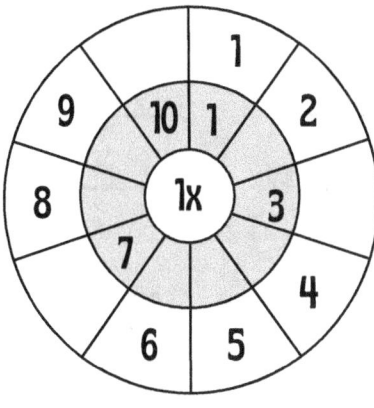
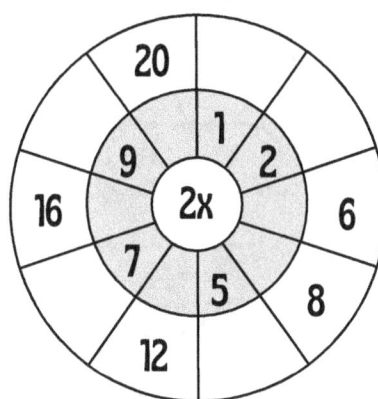
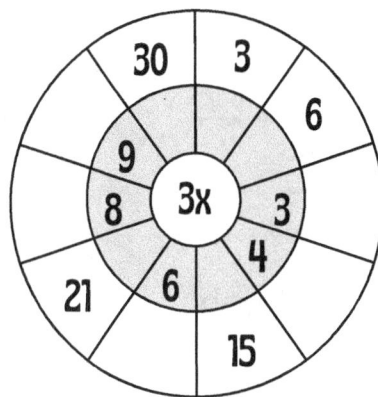
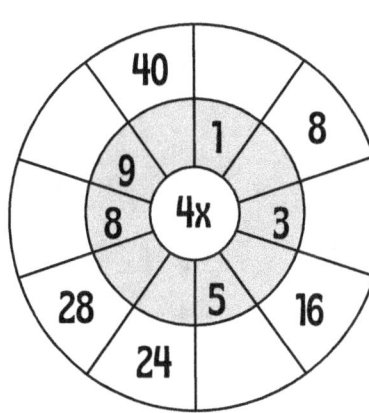
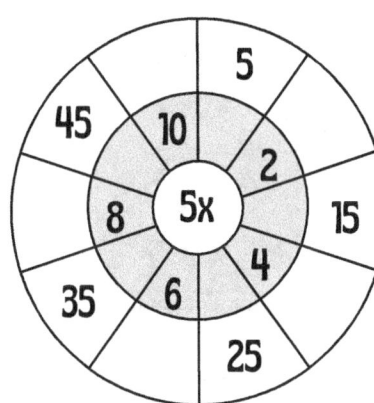
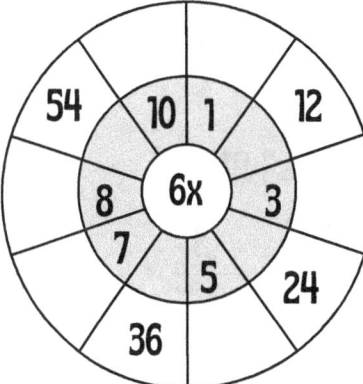
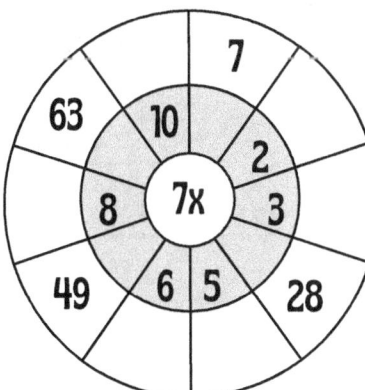
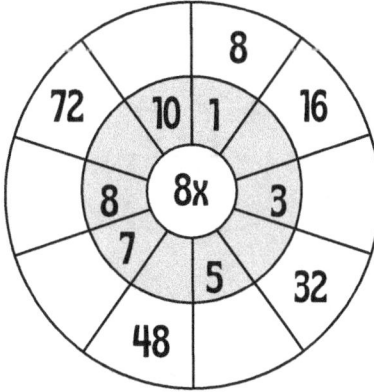
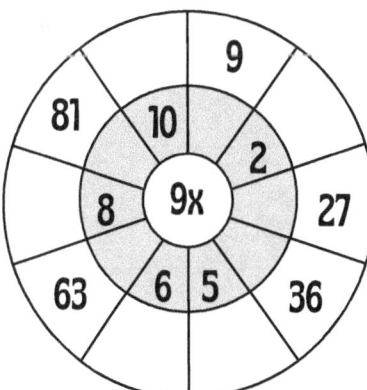

MULTIPLICATION

Solve the following operations

4.648	6.704	12.765	1.634	3.951
× 5	× 3	× 7	× 8	× 6

45.843	5.646	25.180	60.543	72.958
× 2	× 8	× 4	× 9	× 5

65.842	8.406	30.745	3.098	142.329
× 3	× 9	× 7	× 8	× 6

6.907	17.312	3579	21.346	56.212
× 5	× 2	× 4	× 9	× 8

MULTIPLICATION

Solve the following operations

7.459	9.230	2.643
x 23	x 53	x 16

6.208	3.543	1649
x 54	x 83	x 29

5.263	3.938	9.538
x 45	x 74	x 67

MULTIPLICATION

Solve the following operations

| 2.965 × 32 | 4.287 × 70 | 9.034 × 64 |

| 3.737 × 15 | 6.482 × 94 | 8.406 × 83 |

| 56.234 × 32 | 42.462 × 59 | 76.843 × 46 |

DIVISION

Solve the following operations

$9 \div 3 =$

$8 \div 6 =$

$7 \div 2 =$

$5 \div 5 =$

$9 \div 4 =$

$4 \div 2 =$

$6 \div 3 =$

$7 \div 4 =$

DIVISION

Solve the following operations

65 ⌐ 3

121 ⌐ 5

137 ⌐ 6

199 ⌐ 9

DIVISION

Solve the following operations

```
  162 | 7
-  ◯  ◯
  ───
   ◯
-  ◯
  ───
   ◯
```

```
  242 | 5
-  ◯  ◯
  ───
   ◯
-  ◯
  ───
   ◯
```

```
  148 | 4
-  ◯  ◯
  ───
  ◯◯
-  ◯
  ───
   ◯
```

```
  74 | 2
- ◯  ◯
  ───
  ◯◯
- ◯
  ───
   ◯
```

DIVISION

Color the correct answer.

The result of 20 : 4 is...

(6)   (4)   (2)   (5)

The result of 12 : 2 is...

(6)   (5)   (8)   (3)

The result of 24 : 3 is...

(9)   (8)   (6)   (9)

The result of 25 : 5 is...

(8)   (4)   (5)   (6)

DIVISION

Color the correct answer.

The result of 36 : 6 is...

(8) (6) (3) (7)

The result of 40 : 5 is...

(7) (9) (8) (7)

The result of 81 : 9 is...

(9) (5) (8) (7)

The result of 42 : 6 is...

(6) (9) (4) (7)

DIVISION

Color the correct answer.

The result of 42 : 3 is...

(11)   (21)   (14)   (16)

The result of 34 : 2 is...

(14)   (16)   (15)   (17)

The result of 57 : 3 is...

(17)   (19)   (23)   (16)

The result of 64 : 4 is...

(13)   (17)   (16)   (21)

DIVISION

Color the correct answer.

The result of 14:7 is...

(2)   (3)   (14)   (12)

The result of 90 : 5 is...

(16)   (14)   (18)   (15)

The result of 64 : 8 is...

(64)   (8)   (9)   (7)

The result of 126 : 9 is...

(14)   (11)   (10)   (9)

FRACTIONS

Shade the parts to represent the fraction.

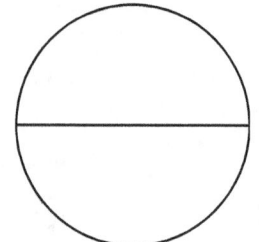 $\frac{1}{2}$ 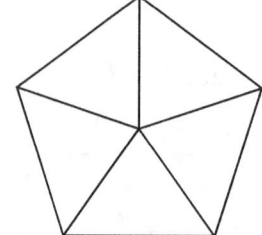 $\frac{2}{5}$

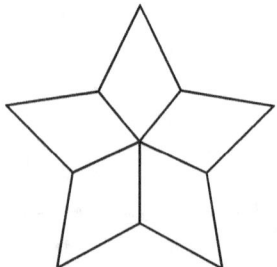 $\frac{1}{5}$ 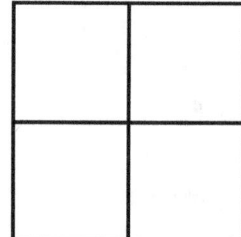 $\frac{3}{4}$

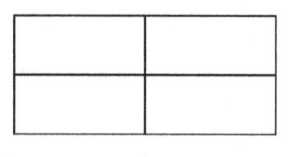 $\frac{1}{2}$ 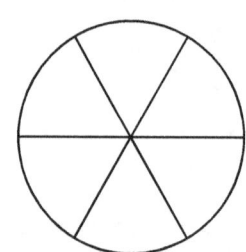 $\frac{5}{6}$

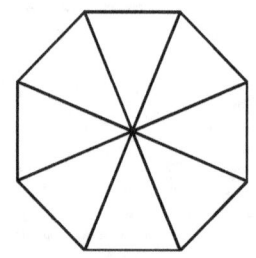 $\frac{3}{8}$ 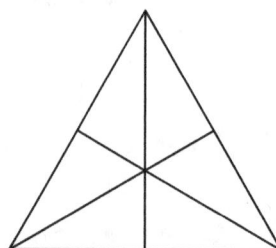 $\frac{1}{6}$

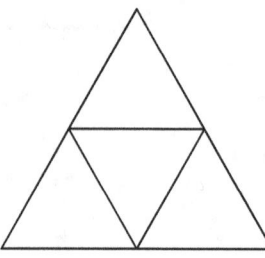 $\frac{1}{4}$ 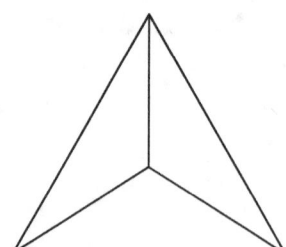 1

FRACTIONS

Write the fraction that each shaded part represents.

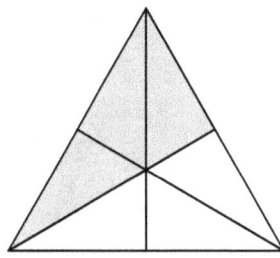  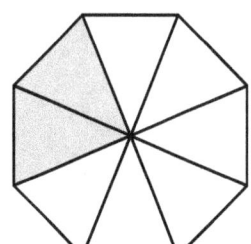 

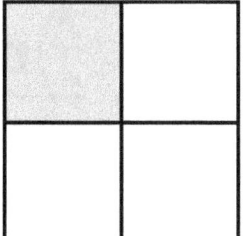  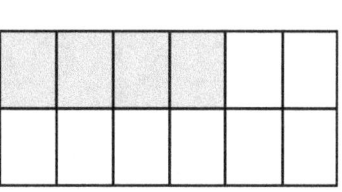 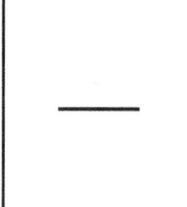

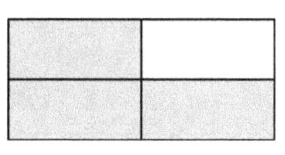  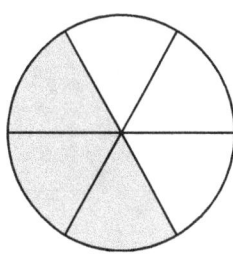 

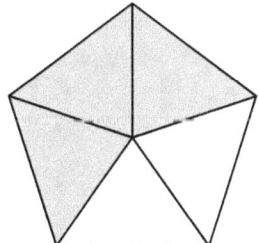  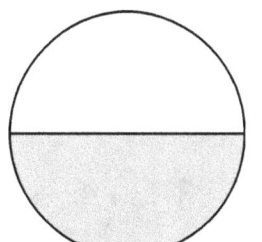 

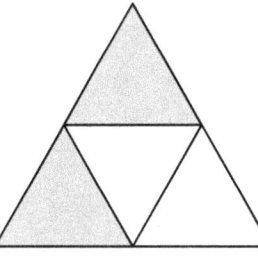  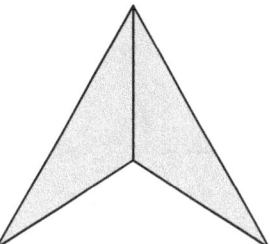

FRACTIONS

Circle the fraction that represents the shaded part.

■□/□□	1/2	3/4	1/4	2/6
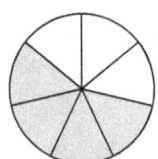	2/4	7/8	3/8	1/8
	4/7	2/3	4/8	3/7
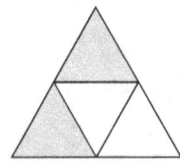	5/5	4/6	1/5	4/5
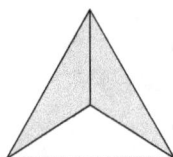	3/4	2/2	1/2	1/4
△	2/3	1/3	3/3	3/2

FRACTIONS

Mark the correct answer with a cross

What fraction is two fifths?

| $\frac{12}{5}$ | $\frac{2}{15}$ | $\frac{2}{15}$ | $\frac{2}{50}$ | $\frac{2}{5}$ |

What fraction is half six?

| $\frac{6}{2}$ | $\frac{6}{12}$ | $\frac{6}{20}$ | $\frac{16}{2}$ | $\frac{6}{3}$ |

What fraction is five eighths?

| $\frac{5}{18}$ | $\frac{15}{8}$ | $\frac{5}{5}$ | $\frac{5}{8}$ | $\frac{8}{5}$ |

What fraction is three quarters?

| $\frac{3}{40}$ | $\frac{3}{4}$ | $\frac{3}{15}$ | $\frac{3}{50}$ | $\frac{3}{40}$ |

FRACTIONS

Mark the correct answer with a cross

What fraction is a third?

| $\frac{3}{1}$ | $\frac{13}{1}$ | $\frac{1}{13}$ | $\frac{1}{3}$ | $\frac{1}{30}$ |

What fraction is eight fifths?

| $\frac{5}{2}$ | $\frac{5}{8}$ | $\frac{8}{5}$ | $\frac{8}{50}$ | $\frac{8}{15}$ |

What fraction is a ninth?

| $\frac{9}{1}$ | $\frac{1}{9}$ | $\frac{19}{1}$ | $\frac{90}{1}$ | $\frac{1}{90}$ |

What fraction is seven tenths?

| $\frac{10}{7}$ | $\frac{10}{10}$ | $\frac{7}{10}$ | $\frac{7}{70}$ | $\frac{7}{7}$ |

PIZZA FRACTIONS

The pizza party was great! How many pizzas were left?
Write your answers as fractions.

COORDINATES

Find and write down the coordinate of each element.

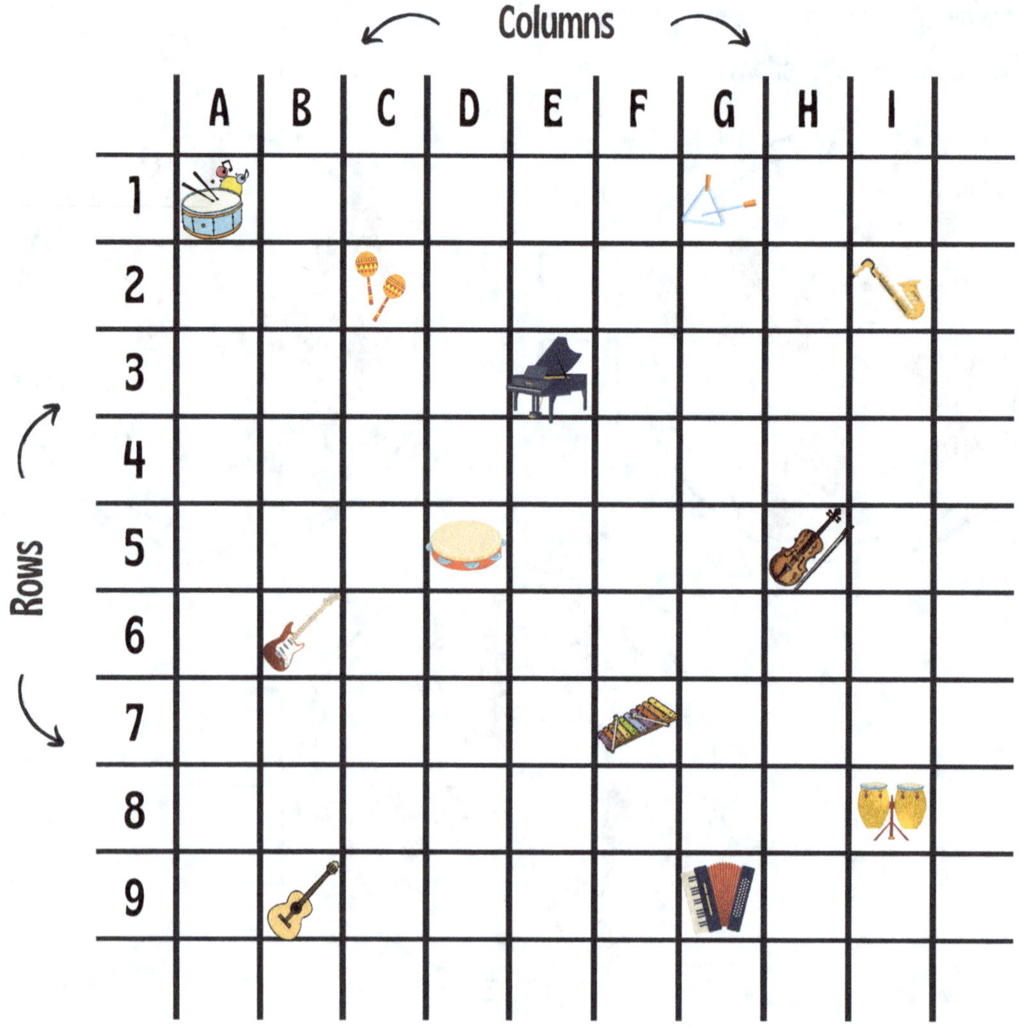

Columns
A
B
C
D
E
F
G
H
I
Rows
1
2
3
4
5
6
7
8
9

(A, 1) (......., ...) (......., ...) (......., ...) (......., ...) (......., ...)

Column Line

(......., ...) (......., ...) (......., ...) (......., ...) (......., ...) (......., ...)

PIRATES COORDINATES

Note the location of each pirate item

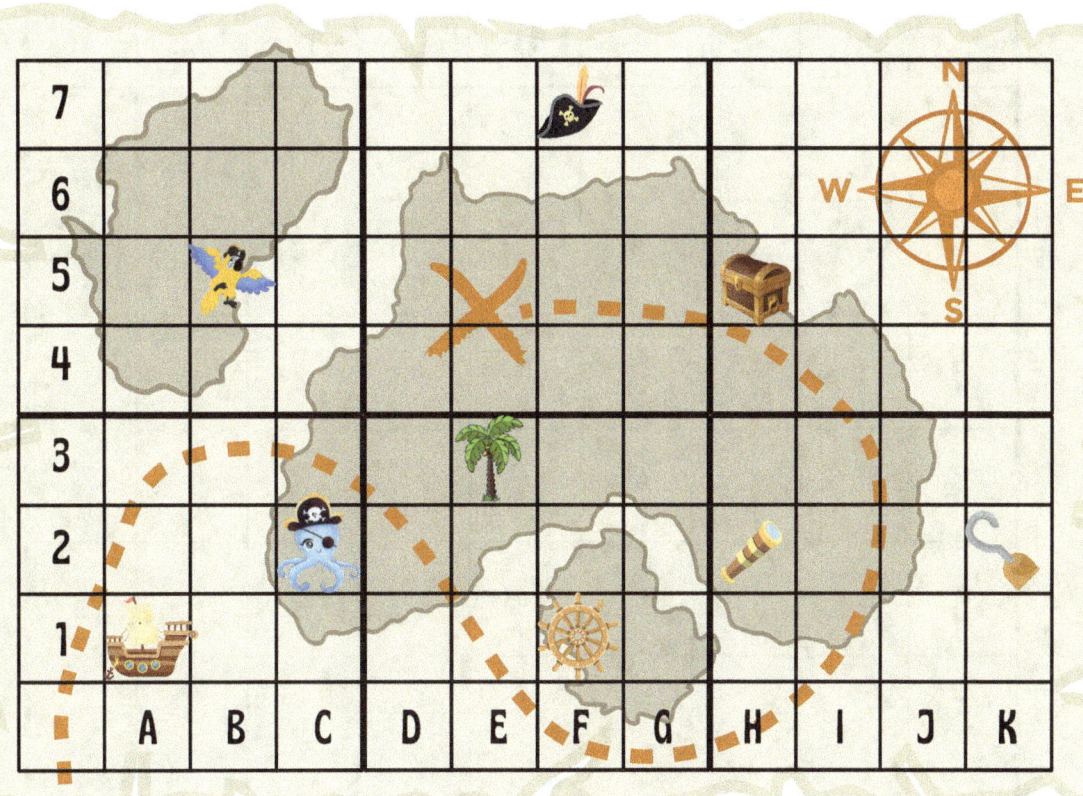

COORDINATES: MAP READING

Describe the characteristics of each coordinate.

	A	B	C	D	E	F	G	H
1								
2								
3								
4								
5								
6								
7								
8								

H1 _____   D3 _____

G6 _____   A8 _____

F2 _____   H4 _____

UNITS OF MEASUREMENT:

TO CONSIDER, READ CAREFULLY

They are used to measure weight, size of objects, length or distance, capacity,

LENGTH

The length of an object or person, the distance between one point and another.

The main unit of length is the meter = m

EQUIVALENCES

1 kilometer = 1,000 meters (m)
1 m = 100 Centimeters (cm)
1 cm = 10 Millimeters (mm)

TIME

The weight that an object or some matter can have.

The unit of measurement for mass is the kilogram (kg)

EQUIVALENCES

1 Kilogram = 1.000 grams (g)
1 g = 1,000 Milligrams (mg)

ABILITY

It is used to measure liquids.

The main unit for measuring capacity is the liter (l).

EQUIVALENCES

1,000 milliliters (ml) = 1 Liter (l)

49

UNITS OF MEASUREMENT: LENGTH

Measure the length of each pencil and write

UNITS OF MEASUREMENT: LENGTH

Measure the length of each pencil and write

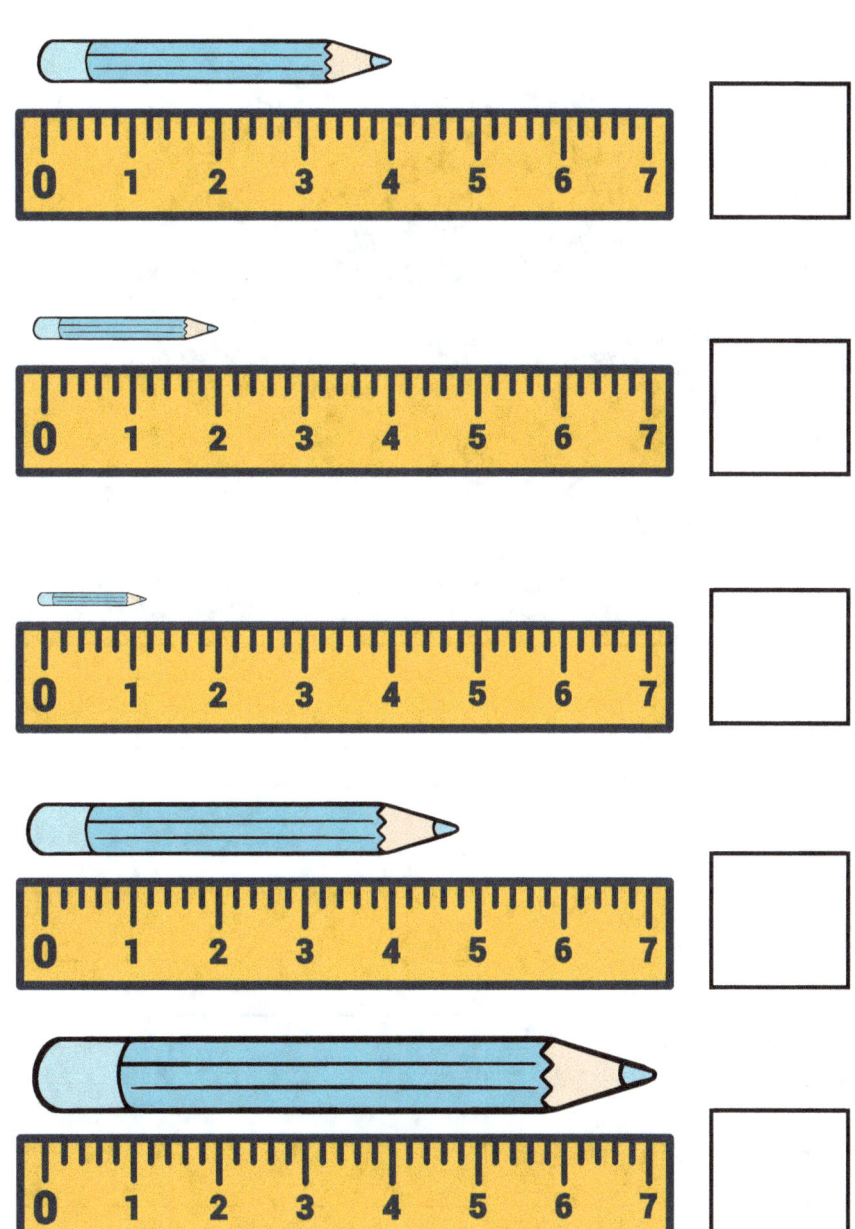

UNITS OF MEASUREMENT: LENGTH

Comparing heights: observe and respond.

Which animal is the tallest?	
Which animal is the shortest?	
How much taller is the elephant than the lion?	
How much shorter is the bear than the giraffe?	
How much taller is the giraffe than the lion?	
How much shorter is the bear than the ostrich?	

UNITS OF MEASUREMENT: LENGTH

Read each problem and answer. Remember to express the correct measurement (kilometers, meters, centimeters)

Luci chased his ball 5 meters on the first throw and 12 meters on the second. What distance did he travel?

Chloe's shirt was 21 inches long. Its sleeves were 15 centimeters long. How much shorter were his sleeves?

Matt and Logan were measuring the fish they caught. Matt's fish was 27 centimeters long and Logan's was 43 centimeters long. How long are the fish together?

The distance from Marta's house to the school is 985 meters. When he has traveled 573 meters he meets his friend Marcos. How many meters do they have left to travel to get to school?

A swan crosses a lake 25 times a day. If you travel 80 m each time, how many kilometers do you travel daily?

A ruler measures 30 cm. If 80 equal rules are placed, one after the other, what length do they occupy? How many meters is it equivalent to?

UNITS OF MEASUREMENT: LENGTH

Observe carefully and respond.

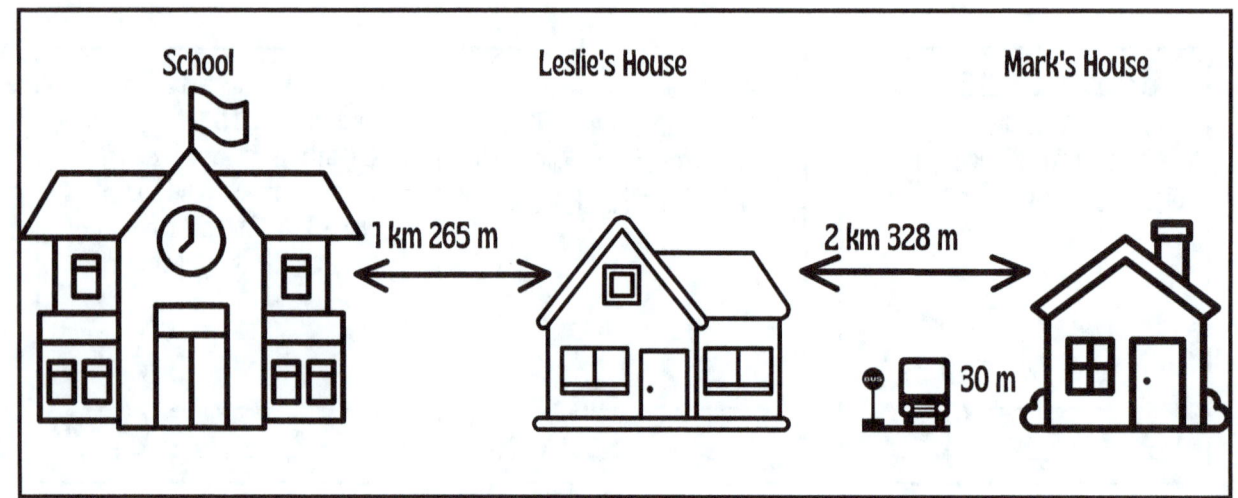

How far is the school from Marcos' house?

If Marcos gets on the bus to go to Leslie's house, how far will he travel?

How far is it from the bus stop to the school?

If the bus makes 3 round trips and 3 return trips throughout the morning from the stop to the school, what distance will it have traveled in total?

UNITS OF MEASUREMENT: LENGTH

HOME CHALLENGE:

With the help of an adult, in your home, locate the following illustrated objects, measure their length as indicated and write.

Remember to express the correct measurement: m, cm

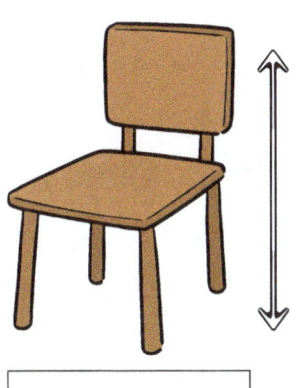

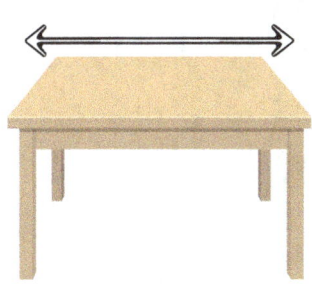

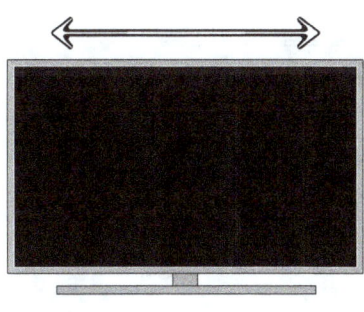

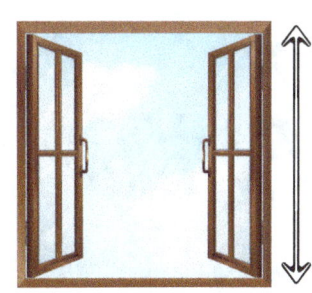

UNITS OF MEASUREMENT: LENGTH

HOME CHALLENGE:

With the help of an adult, in your home, locate the following illustrated objects, measure their length as indicated and write.

Remember to express the correct measurement: m, cm

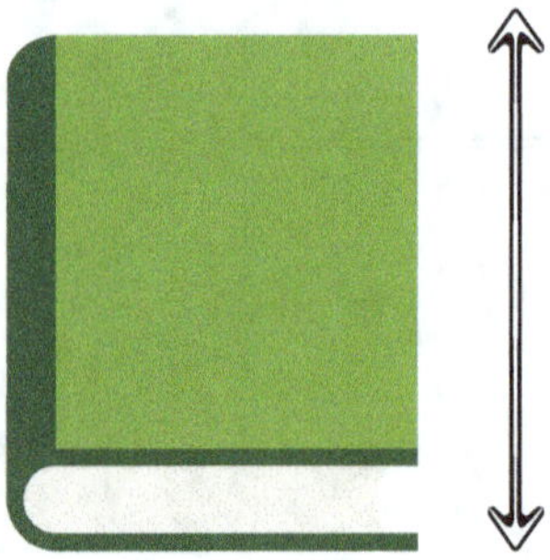

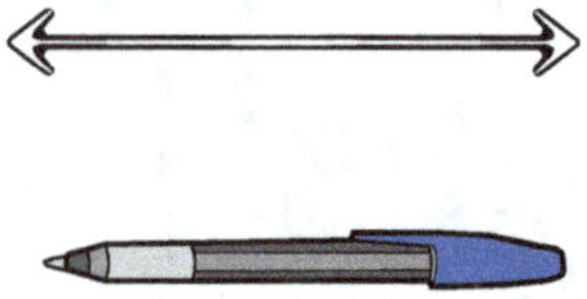

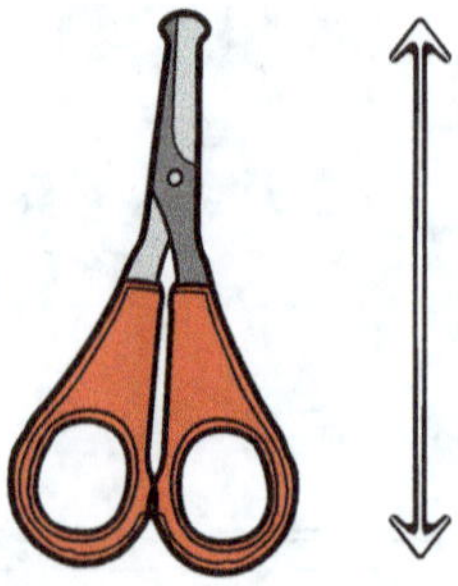

UNITS OF MEASUREMENT: TIME

Cut and paste each element where it corresponds according to its weight.

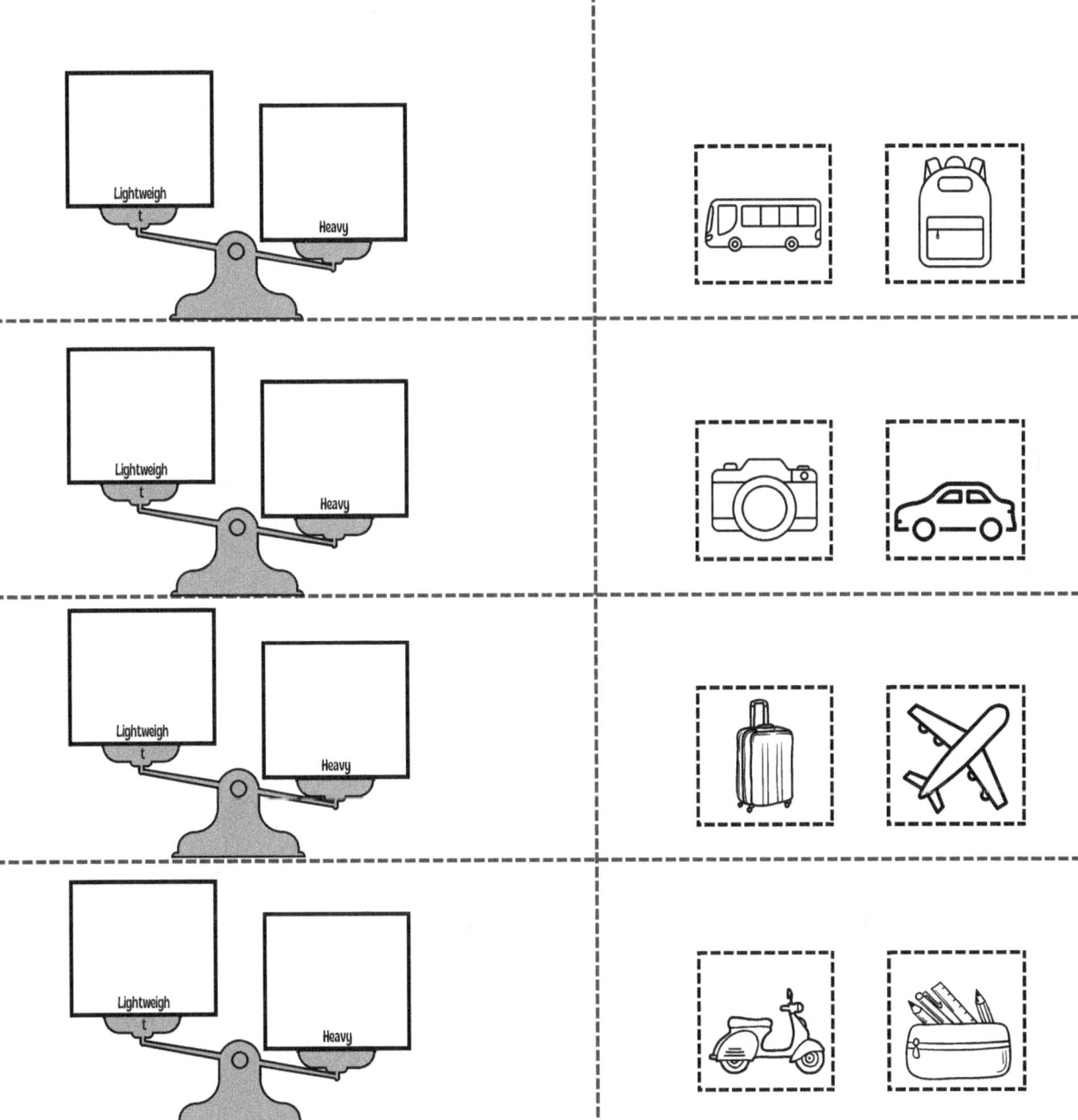

UNITS OF MEASUREMENT: TIME

How much weight is there? Match with the correct answer.

Keep in mind that: 1 kg = 2 half kilos = 4 quarter kilos

kilo + Half a kilogram + quarter kilo + quarter kilo	1 and a half kilo
quarter kilo + quarter kilo + quarter kilo + quarter kilo + Half a kilogram	3 and a half kilos
kilo + kilo + quarter kilo + quarter kilo + Half a kilogram	1 kilo and a quarter
Half a kilogram + Half a kilogram + quarter kilo	2 kilos
kilo + kilo + Half a kilogram + Half a kilogram + Half a kilogram	3 kilos

UNITS OF MEASUREMENT: TIME

Grams or kilograms? Select the correct answer.

150 g   5 kg

450 g   15 kg

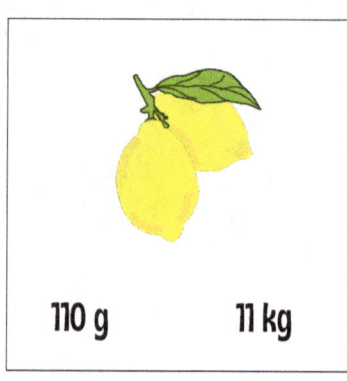
110 g   11 kg

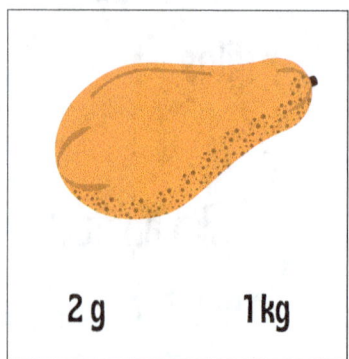
2 g   1 kg

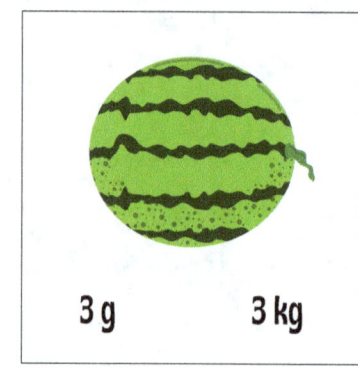
3 g   3 kg

10 g   10 kg

80 g   8 kg

60 g   6 kg

750 g   10 kg

UNITS OF MEASUREMENT: TIME

Check true if the following is true, check false if it is false.

TRUE ◯ ◯ False	TRUE ◯ ◯ False
TRUE ◯ ◯ False	TRUE ◯ ◯ False
TRUE ◯ ◯ False	TRUE ◯ ◯ False

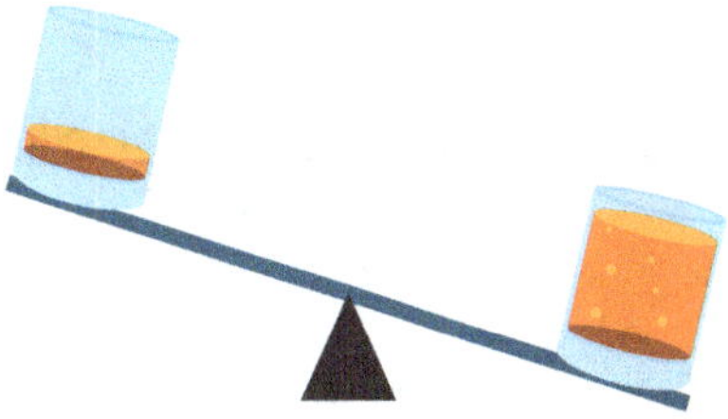

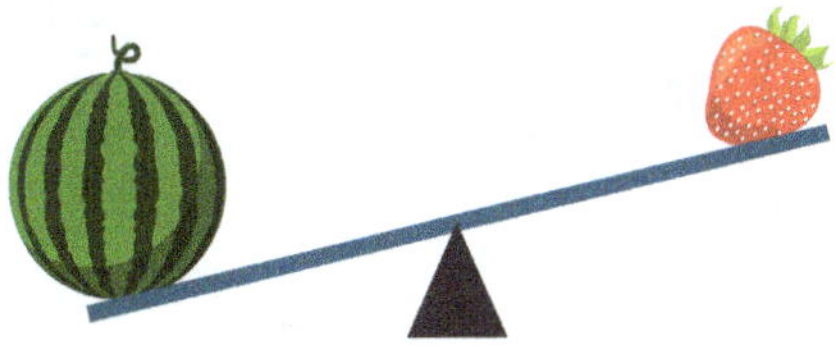

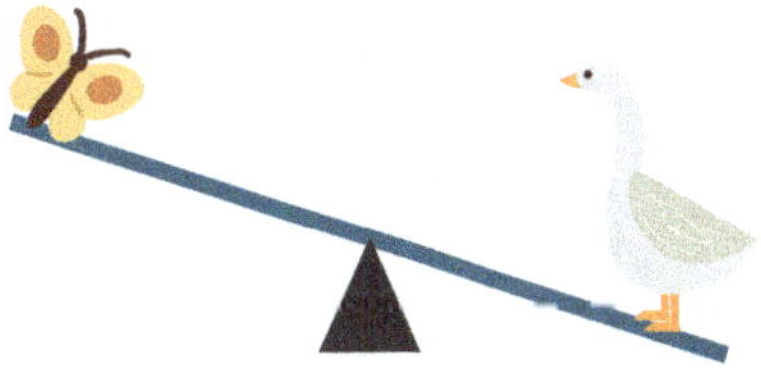

TRUE ◯   ◯ FALSE

UNITS OF MEASUREMENT: TIME

Get ready for an epic excursion!
Pack only what you really need.

Circle the items you can bring that, added together, weigh exactly 1 kg.

UNITS OF MEASUREMENT: TIME

Write the equivalences of the units of weight.

Remember: 1 kg = 2 half kilos = 4 quarters of a kilo

12 quarters of a kilo is equivalent to .... half a kilo.

6 kilos is equivalent to .... quarters of a kilo.

8 quarters of a kilo is equivalent to .... kilos.

10 kilos is equivalent to .... quarters of a kilo.

4 kilos is equivalent to .... half kilos.

10 quarters of a kilo is equivalent to .... half a kilo.

2 half kilos is equivalent to .... kilos.

6 kilos is equivalent to .... quarters of a kilo.

12 kilos is equivalent to .... quarters of a kilo.

UNITS OF MEASUREMENT: TIME

HOME CHALLENGE:
With the help of an adult, at home, use the scale and compare the weight of 4 objects or fruits. Remember to express the correct measurement: g, k.

Object 1............................. Object 2.............................

Object 3............................. Object 4.............................

Draw and order the previous objects or fruits from greatest to least weight.

UNITS OF MEASUREMENT: ABILITY

Circle the appropriate measurement for each object.

250ml   25 liters

1000ml   100 liters

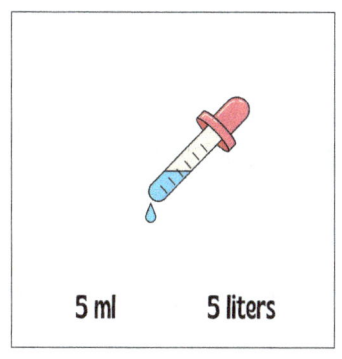

5 ml   5 liters

1 milliliter   1 liter

100ml   10 liters

500ml   5 liters

5 liters   500ml

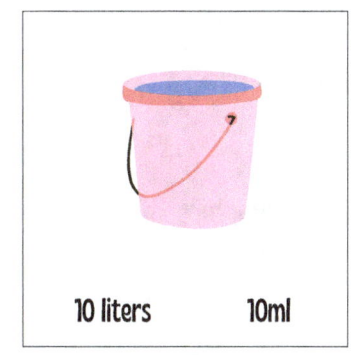

10 liters   10ml

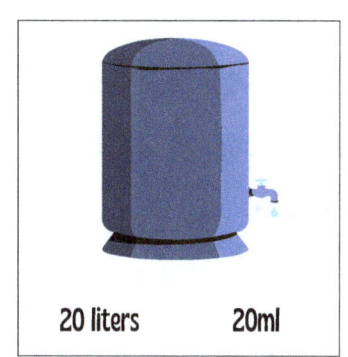

20 liters   20ml

UNITS OF MEASUREMENT: ABILITY

Write the equivalences between the measures of capacity

Remember: 1 l = 2 half liters = 4 quarters of a liter

1 liter is equivalent to .... half a liter.

2 quarts of a liter is equivalent to .... half a liter.

16 quarts of liter is equivalent to .... liters

9 liters is equivalent to .... half liters.

4 liters is equivalent to .... quarters of a liter.

9 half liters is equivalent to .... quarters of a liter.

8 quarts of a liter is equivalent to .... liters.

10 half liters is equivalent to .... liters.

12 liters is equivalent to .... quarters of a liter.

UNITS OF MEASUREMENT: ABILITY

Read each statement carefully and respond.

How many glasses of milk can be filled with the bottle that contains 2 liters?

Answer

How many 1/2 liter jugs can be filled with water from the bottle?

Answer

How many glasses of soda can be filled with the liquid from the can?

Answer

UNITS OF MEASUREMENT: ABILITY

Read each statement carefully and respond.

How many 1/2 liter cups can be filled with the liquid from the container?

Answer

How many 1/2 liter cups can I fill the bucket with?

Answer

How many 2L bottles will it take to fill the fish tank with water?

Answer

UNITS OF MEASUREMENT: ABILITY

Read each statement carefully and respond.

Remember 1000 ml = 1 l.

In the kitchen, Mom has a jug with 800 ml of orange juice and another with 450 ml of apple juice. How many liters of juice does mom have in total?

Juan had a liter of water in a jug and poured 250 ml to make a glass of milk. How many milliliters of water are left in the jug?

If we fill a half-liter glass with water, how many milliliters of water have we used?

Ana is preparing a punch. You need half a liter of orange juice, a quarter of a liter of lemon juice and a quarter of a liter of pineapple juice. How many liters of juice will you need in total for your punch?

UNITS OF MEASUREMENT: ABILITY

Read each statement carefully and respond.
Remember 1000 ml = 1 l.

In the refrigerator there is a container with 2 liters of pear juice and another container with 1 and a half liters of apple juice. If mom takes out half a liter of pear juice and adds a quarter of a liter of orange juice, how many liters of juice are there now in total in the refrigerator?

Orange juice comes in 750 ml containers and pineapple juice in 500 ml containers. If we buy 2 containers of orange juice and 3 containers of pineapple juice, how many milliliters of juice in total do we have?

If a child needs 150 ml of cough syrup per day, how many milliliters will he need for the entire week?

If a bottle of juice has 2 liters and we want to divide it into 8 equal parts, how many milliliters will there be in each part?

ORDINAL NUMBERS

TO CONSIDER, READ CAREFULLY

We use ordinal numbers for dates or the order of something.

1° FIRST	11° ELEVENTH	20 TWENTIETH
2° SECOND	12° TWELFTH	30 THIRTIETH
3° THIRD	13° THIRTEENTH	40° FORTIETH
4° ROOM	14° FOURTEENTH	50° FIFTIETH
5° FIFTH	15° FIFTEENTH	60° SEXAGÉSIMO
6° SIXTH	16° SIXTEENTH	70° SEVENTIETH
7° SEVENTH	17° SEVENTEENTH	80° OCTOGÉSIMO
8° OCTAVO	18° EIGHTEENTH	90° NONAGÉSIMO
9° NINETH	19° NINETEENTH	100° HUNDREDTH
10° TENTH		

ORDINAL NUMBERS

Write below each runner the correct order of their position.

ORDINAL NUMBERS

Identify the position of the children starting from the left. Write in symbols.

Carlos   Zed   Marcos   Leo   Kate

John   Caro   Karla   Ben   Well

John: SIXTH/ 6TH

Zed:

Ben:

Kate:

Caro:

Marcos:

Carlos:

Leo:

And:

Male:

ORDINAL NUMBERS

Relates to its corresponding

4° • • Fifth

8° • • Third

2° • • Eighth

6° • • Seventh

7° • • Fourth

3° • • Second

10° • • First

9° • • Ninth

1° • • Sixth

5° • • Tenth

ORDINAL NUMBERS

Relates to its corresponding

17° •   • TWENTY

11° •   • EIGHTEEN

20° •   • ELEVENTH

15° •   • TWELFTH

19° •   • THIRTEENTH

13° •   • SEVENTEENTH

12° •   • FIFTEENTH

18° •   • FOURTEENTH

14° •   • FIFTEENTH

15° •   • NINETEENTH

ORDINAL NUMBERS

Write the ordinal numbers in words

7°

12°

20°

5°

27°

5°

40°

11°

18°

22°

ORDINAL NUMBERS

Place a V if the statement is true and an F if it is false.

🐰	☐ The mouse is in fifth place.
🐢	☐ The elephant is in eighth place.
🐌	☐ The hedgehog is in eleventh place.
🐸	☐ The crocodile is in eleventh place.
🦆	☐ The duck is in fifth place.
🐭	☐ The snail is in fourth place.
🐘	☐ The rabbit is in first place.
🐿️	☐ The fox is in tenth place.
🦔	☐ The turtle is in second place.
🦊	☐ The frog is in fourth place.
🐊	☐ The squirrel is in eighth place.

Dear readers,

Thank you for embarking on this exciting mathematical logic adventure with me! It has been a pleasure to create this workbook designed especially for curious and creative minds.

Support from readers like you is invaluable to indie authors like me. If you have a moment, I would greatly appreciate it if you could leave a review on Amazon. Your opinions are crucial so that more parents, teachers or adults who love early childhood education find these exercises.

I invite you to visit my page on Amazon and by following me you can access the different books that I am preparing for our children.

Thank you for contributing to the reading community and making it possible for more people to discover this book!

With gratitude, Mariledys

Scan to leave your comment or visit my page on Amazon

Scan to receive free information and resources or if you need to access solution material or answers in this book.

You can also write to
mariledys@educkidsonline.com

www.ingramcontent.com/pod-product-compliance
Lightning Source LLC
Chambersburg PA
CBHW062227220526
45471CB00009B/3373